By Braden Perry Clark

Publisher, Author & Illustrator: Braden Perry Clark
Editors: Tondee Perry Clark & Camille Clark
Cover and Header Font:
Agent Orange available at pizzadude.dk
ISBN# 979-8-9909277-1-1

Makeover: The First April Fools' Joke: While the T-Rex was asleep, two ancient mammals, Didelphodon and Meniscoessus, painted him the colors of a prehistoric purple dinosaur! The passing Triceratops noticed something strange - but funny!

Grand Canyon Roommates: In the Grand Canyon during the ice age, lived Northrotheriops - a giant sloth which was the size of a black bear. It inhabited caves and wasn't a clean animal. In the caves, you could find evidence of what it ate (aka dung!). It wasn't the only animal that liked to live in the caves, the Miracinonyx is similar to a cheetah except that it had the lifestyle of a snow leopard. It hunted goats and brought them to the caves. I imagined the scenario when both animals met.

Godzilla Hates Noise: The reason why Godzilla attacks cities is that he CAN'T STAND THE NOISE!

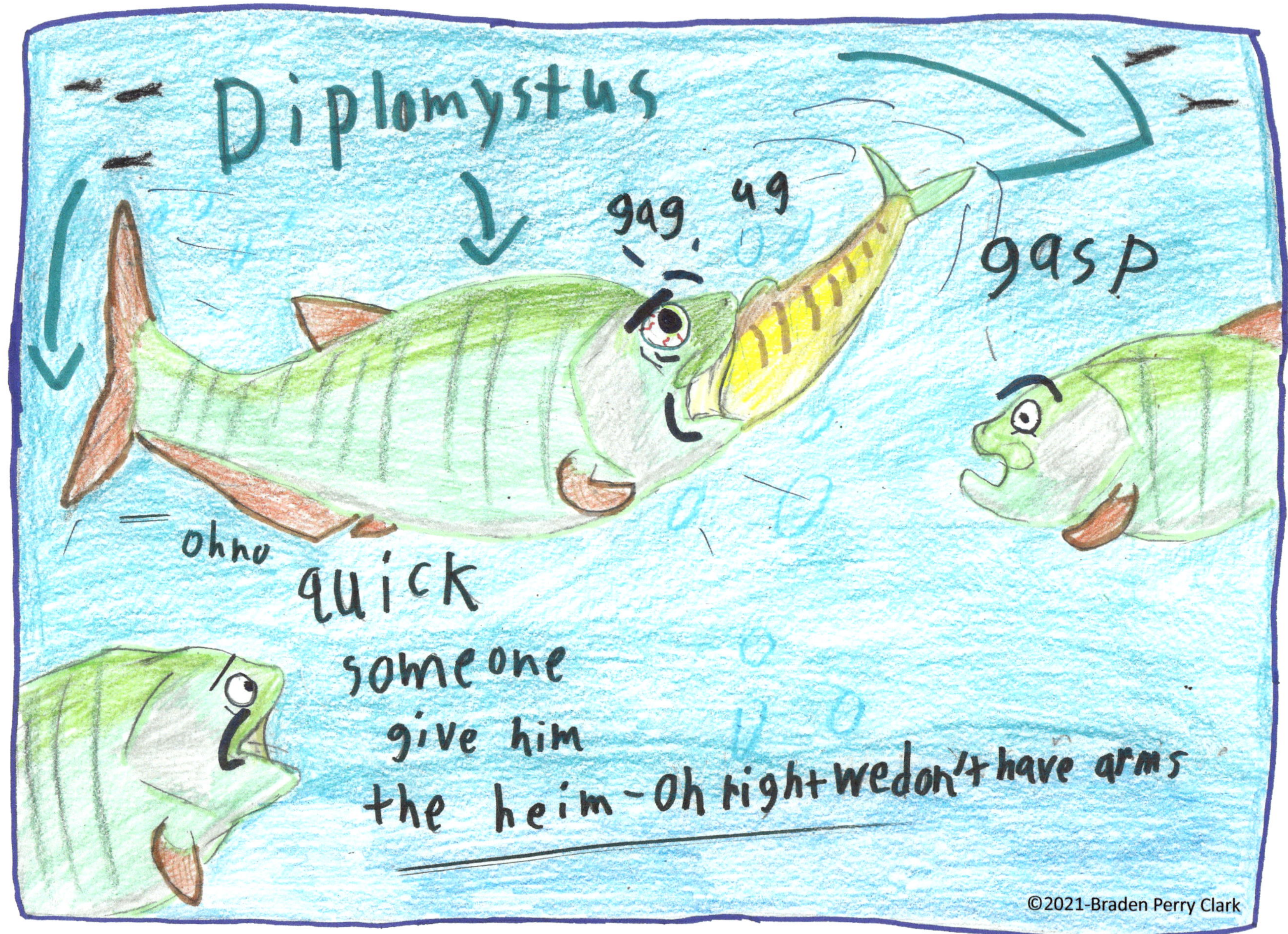

Heimlich: This poor fellow is Diplomystus, a fossil fish from the Green River, Wyoming formation. He thought more with his stomach than his brain and you can see how that didn't work out so well for him. His fossil was found with his meal still in the mouth. He's not the only fossil fish to suffer such a fate. Too bad there isn't a Heimlich maneuver for fish!

Hosed Awake: Here's my story which takes place in Texas during the Middle Miocene period. A herd of Gomphotherium Productum was on their way to a local pond when they came across a pack of sleeping Aelurodon which are bone-crushing dogs. That's when one of them got the bright idea of grabbing some water from the pond and bringing it over to hose the Aelurodons awake! Let's just say that these creatures won't be napping out in the open anymore! Fact: Gomphotherium are relatives to modern-day elephants. However, their lower jaws are bigger.

Children's Artwork on Crest: This sleeping Dad is a Pterosaur called Tupandactylus Imperator that lived in the early Cretaceous Period and was found in the Crato Formation of Northeastern Brazil. The crest on his head is the largest of its kind and is the perfect place for the kids to leave their artwork!

Skunk in La Brea Tarpits: What can be worse than getting stuck in the La Brea Tarpits? Getting stuck with a skunk!

Danger on Both Sides: Time and Location: Late Miocene Period in Northeastern Argentina. The family of Protypotherium got themselves caught between a Macroeuphractus - a fully carnivorous armadillo and a terror bird named Andalgalornis. This scene was created based on badger hunting behaviors. They burrow to find their prey. If any animals try to escape out of the burrow, a coyote will be waiting for them. This scene is speculative but probably could have happened. Fact: Protypotherium is a Litoptern, not a rodent.

Very First Parasites: This poor brachiopod (Neobolus Wulongqingensis), a shellfish that has a bad infestation of parasitic tube worms. These guys steal the food he is trying to eat. Plus, they grow on his shell which makes it hard for him to eat and grow - like a bad roommate!

Dumb Mammoth Moments: Male mammoths often got into trouble due to inexperience when they left their mother's herd and lived on their own. Normally, their mothers would get them out of trouble or steer them away from dangerous situations. They taught them the basics of life. However, being males, they probably felt that they were invincible and since no other animal wanted to challenge them, they often did stupid things. It's a shame that their mother's wisdom didn't stick. This picture illustrates a silly situation that shows the male mammoth brain and its ability to not think things through.

360 Degree Vision: Meet the fossil that was discovered recently. It is the Palaeotanyrhina Exophthalma. The fossil was found in Burmese Amber and dated back 100 million years. It has 360-degree vision, thanks to its big eyes that stick out. The vision capabilities are useful for finding food, avoiding predators, or babysitting mischievous nymphs. Fun fact: this insect is in the group that includes bed bugs.

Motorcycle: Collinsium Ciliosum also known as the "Spikey Worm". Back when life was very young, this creature was a worm with legs and spikes. I picture this guy as a motorcycle gang member who rides his motorcycle under the sea wearing a cool leather jacket.

Outhouse Terror: Location: Egypt; Time Period: Eocene; A Moeritherium, an ancient relative of the elephant, falls victim to a 22-foot-long Gigantophis (a giant snake) in the local outhouse. Unfortunately, the next guy in line will have a long wait!

Scratching the Dirt: This scene is based on fossilized tracks found in Roubideau Creek in Delta County, Colorado. It captures the dating dance of probably the Acrocanthosaurus. The exact reason for the scratching is unknown, it's probably to show that the male can build a nest for future offspring. These tracks are an interesting insight into theropod behavior - although it's probably weird to witness the activity.

©2021-Braden Perry Clark

Secret Side of Herbivores: Triceratops are normally herbivores, but occasionally, they might eat a little meat like this guilty guy who has been joining his T-Rex friend for some BBQ.

©2021-Braden Perry Clark

The Cheater: Bistahieversor made the rule that his friend, Pentaceratops, couldn't use the frill at the top of his head to help him in the game. However, he never listened. Note: Bistahieversor is a relative of T-Rex and Pentaceratops is a relative of Triceratops.

The Big Angry Dad: Place: Mongolia in the Iren Dabasu Formation. Time: Late Cretaceous Period. Who: An Alectrosaurus (a type of Tyrannosauroid) is out looking out for something to eat when it stumbles upon a nest of giant eggs. However, before he can even touch the eggs, the father of the eggs. Gigantoraptor, a 26-foot-long Oviraptorosaur is not happy to see an intruder in its nest. Let's just say that the small Alectrosaurus is going to be taught a lesson.

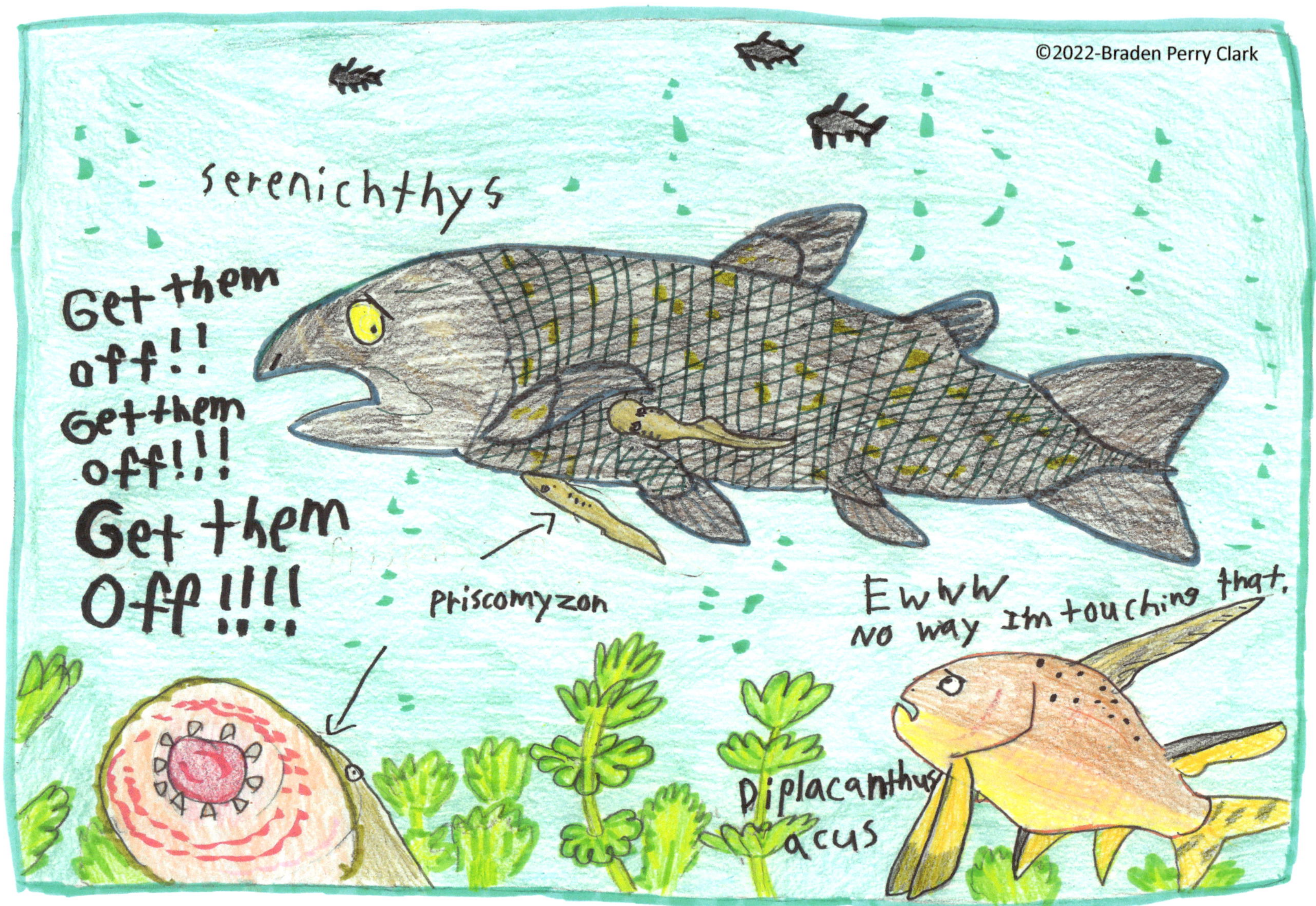

Vampire Fish: This poor Serenichthys is being preyed upon by the oldest known lamprey called Priscomyzon - a real-life vampire fish. Fossils of the late Devonian period were found at Waterloo Farm in South Africa showing these creatures. The bottom drawing shows a close-up of the mouth of the Priscomyzon and shows that it is like modern-day lampreys. This fish will have a hard time getting these slimy bloodsuckers off and will receive no help from any other fish.

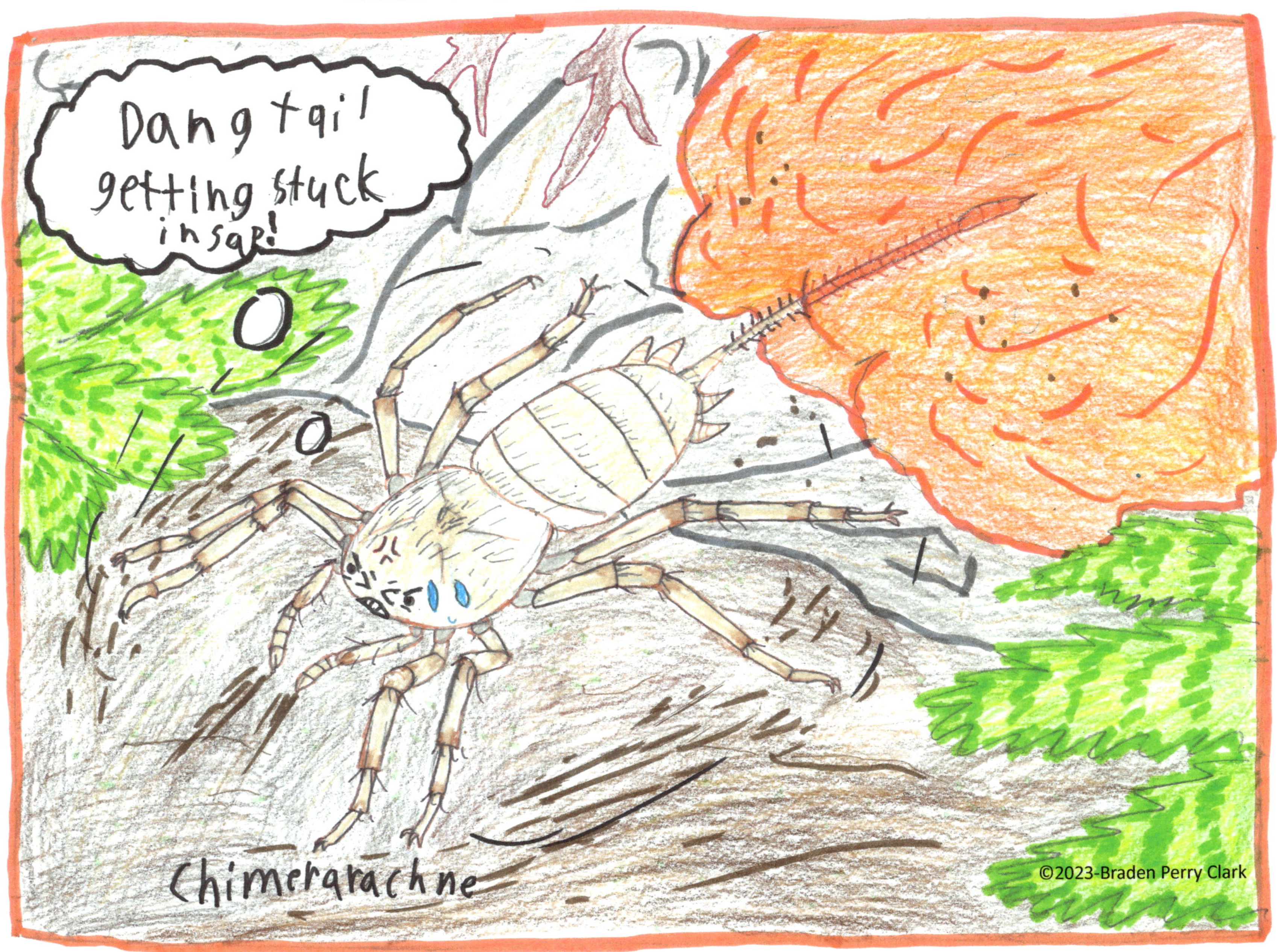

Why Spiders Don't Have Tails: This unfortunate fellow stuck in sap is called the Chimerarachne. It is an ancient spider that had a tail. It lived in Myanmar during the Cretaceous Period. A fossil was found in the Burmese Amber. Spiders today aren't known to have tails, probably because it gets in the way of its web-making.

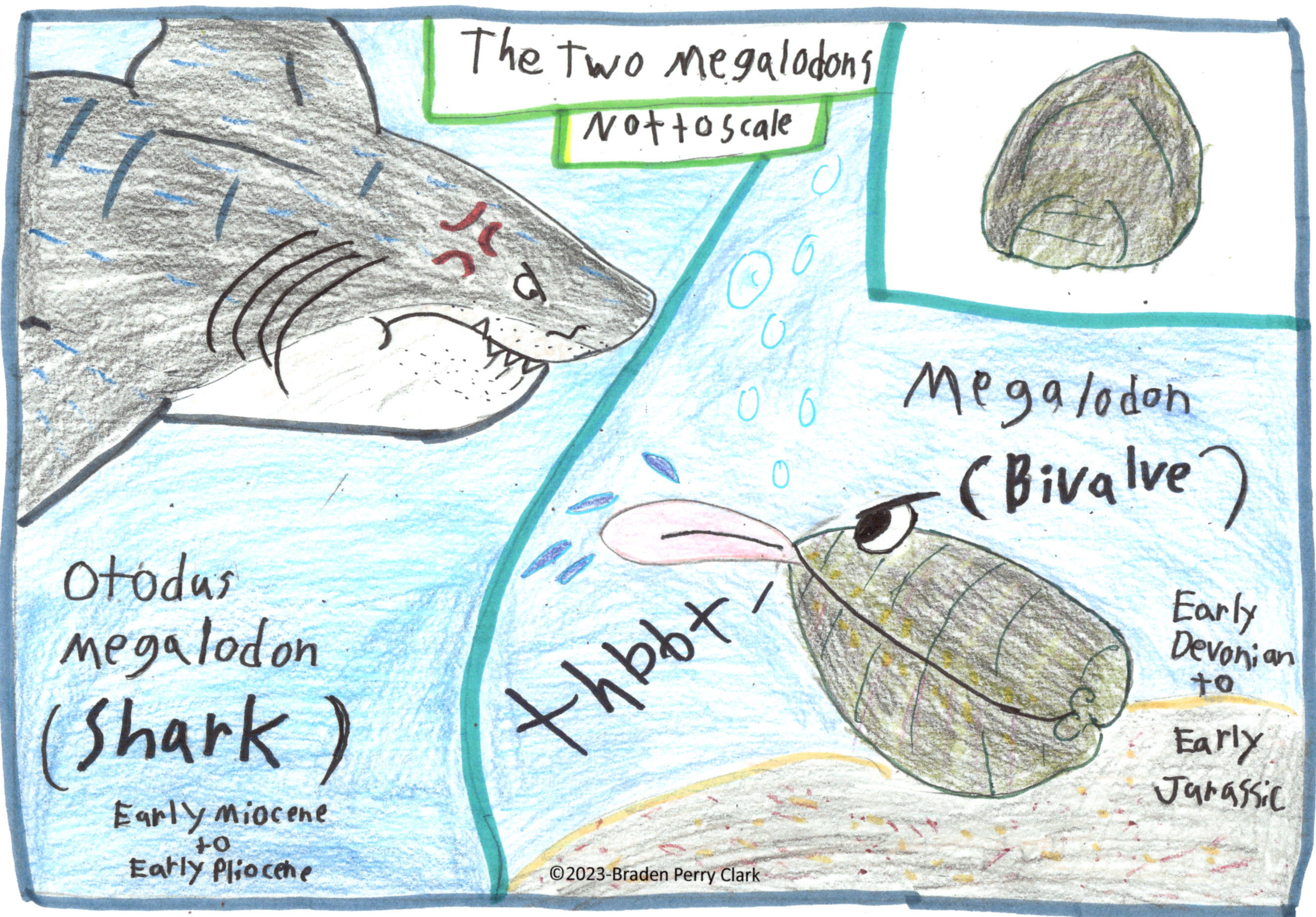

Who is the real Megalodon? You might have heard of the famous Megalodon Shark. You might be interested to learn that there is a Bivalve (a mollusk) also named Megalodon. In my drawing, the shark is annoyed at the Bivalve for having the same name as he, and the Bivalve, in a defiant act, gives a raspberry! In fact, the Bivalve family group it is from, is named Megalodontidae. The name Megalodon is a popular name and shows up in some other groups of animals. Talk about confusing!

Baseball: A Stegosaurus decides to join the Camptosauruses for a game of baseball. However, there's a reason why spiky dinosaurs don't play sports - the ball sticks to their spikes! (Hint - look at the tail!)

Bigfoot: With drones flying around, it's harder for Bigfoot to stay hidden.

Dolly's Illness: About 150 million years ago in the Morrison Formation in southwest Montana, the Diplodocus - nicknamed "Dolly" was suffering from an illness that caused it to cough a lot. The illness was so bad it left a mark on its neck bones where the air sacks were. The other creatures around Dolly tried to keep their distance.

Godzilla Facts: Before the battle, Godzilla states a few facts to King Kong, "Listen here, Kong. I want to let you know that I have a lot of abilities - from atomic breath to nuclear pulse and regenerative powers. I have fought many monsters, aliens, robots, and many more enemies - some more powerful than you! I have been through many things - from getting nuked; getting hit by an asteroid; falling into a volcano and surviving a man-made black hole (2 times!). So, what have you accomplished, APE?"

Spiky Egg with A Mouth: This fellow is Saccorrytus, a small organism that lived in the ocean in the early Cambrian Period. It is distinct for having a big mouth and spikes all over it. This scene portrays this organism dressed as a clown handing out balloons at a kid's birthday party. Unfortunately, he got fired after this event.

Similar situations different timelines: the top picture show Synapsids (not Dinosaurs) during the Permian Period; the bottom picture is modern time. In both situations, the prey is trying to escape the predator and hide in the tree.

Horseplay: Two Hagerman Horses pranked the Borophagus - Bone Crushing Dog - by creating a fake horse made of hay. They are laughing at him and enjoying their prank.

The Misunderstood: The Cotylorhynchus - an animal that used to live in North America before the dinosaurs. Its small head compared to its large body makes it look like it's swollen.

Spiked Behind: Late Jurassic Period. Tanzania. Somewhere in the Tendaguru Formation, a Giraffatitan is playing Frisbee with his friends on the beach. He was having so much fun until he stumbled backward and accidentally fell on a Kentrosaurus who was enjoying quiet time on the beach. AAAAAAAAAHHHHHHHHHHHH! Note: An adult Giraffatitan could grow to 39 feet while a Kentrosaurus could only grow to about 14 feet.

Toothy Shell: Fast fact: This drawing is about the group of jawless fish called Ostracoderm that lived during most of the Paleozoic Era. Their shells are made of the same substance as teeth. The material fused together to form a hard shield of armor on their heads. This underwater scene in the ocean imagines a fish shower. A bottle of shell paste (kind of like toothpaste) is handy for the jawless fish who wants a shiny shell.

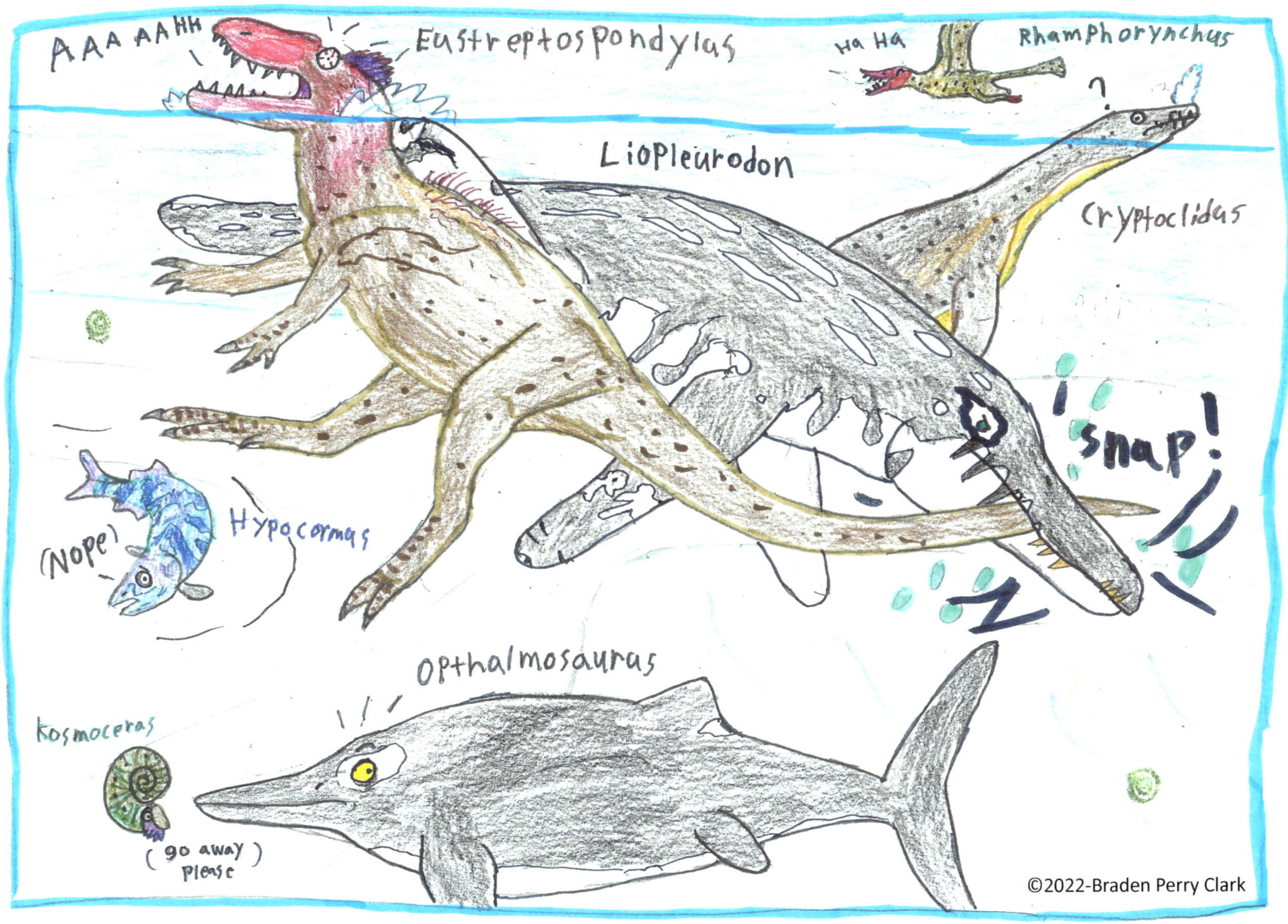

Cruel Sea: This drawing is based on an episode of Walking With Dinosaurs called "Cruel Sea". I wanted to give an updated version of the animals that were shown in that episode. This picture depicts an Eustreptospondylus that was washed out to sea during a storm and is now trying to swim back to shore, but it was caught by a Liopleurodon. The Liopleurodon was once thought to be a massive 82-foot creature. However it has now been determined to be only 23 feet long. New research has found that Ophthalmosaurus has darker skin color than once thought.

Theme Park Riot: "I knew this sign on an island of dwarfs is a bad idea," said the Ampelosaurus who was stationed at the entrance of the new theme park. The natives of the island, mostly Paludititan and Magyarosaurus - small Sauropods - were at first excited about the new theme park. But when they arrived at the entrance, they were outraged that they didn't meet the height requirement! So, they started rioting!

Fish Revenge: When a hungry Spinosaurus caught a fish, he thought he was very lucky until a tree fell on him with the help of Onchopristis (a giant sawfish).

Caecilian Infiltrator: The worms are having their party somewhere underground in the tropics. They were having a great time but started to notice that some of their friends were missing. Little did they know, their party had been infiltrated by a Caecilian - a type of amphibian that looks like a worm but has a skeleton and a double row of teeth. The infiltrator was enjoying the party, especially the refreshments!

©2021-Braden Perry Clark

No Jaw No Service: A jawless fish wanted to get a new type of food that is all the rage. But when he got to the restaurant, they only served fish with jaws and he's a jawless fish! This takes place before the dinosaurs. Why can't everyone be nice to one another? Even animals?

Mobster: I imagine that back in the Cretaceous Period in the Burmese Amber Area, small critters had to worry about meeting their loan payments to the criminal organization known as 'The Resins!" They earned their name because of how they punished anyone who didn't pay off their loans by putting them in the tree resin (that would later harden into the amber we currently know). They were put on display as a warning to others."

Crown Transfer: Brachyopoidea was once the king of the freshwater systems throughout the world. Crocodylomorph stole the crown by walking up behind the Brachyopoidea and kicking him to down under - literally. The species survived for a while in Australia and Antarctica during the Jurassic and Cretaceous Periods but never regained their former glory.

Godzilla School: Godzilla teaches his sons that human civilization is bad, and nature is good.

Intruder in the Bathroom: Don't you hate it when you have a bug wander into your bathroom? Well, during the Carboniferous Period, there was a 27.6-inch scorpion named Pulmonoscorpius! This Silvanerpeton went a little pale when he saw this intruder in his bathroom!

Unfair Crabby: Carl the False Crab was denied access to the tide pool because only true crabs are allowed to enter. Fact: False crabs are more closely related to lobsters than True Crabs.

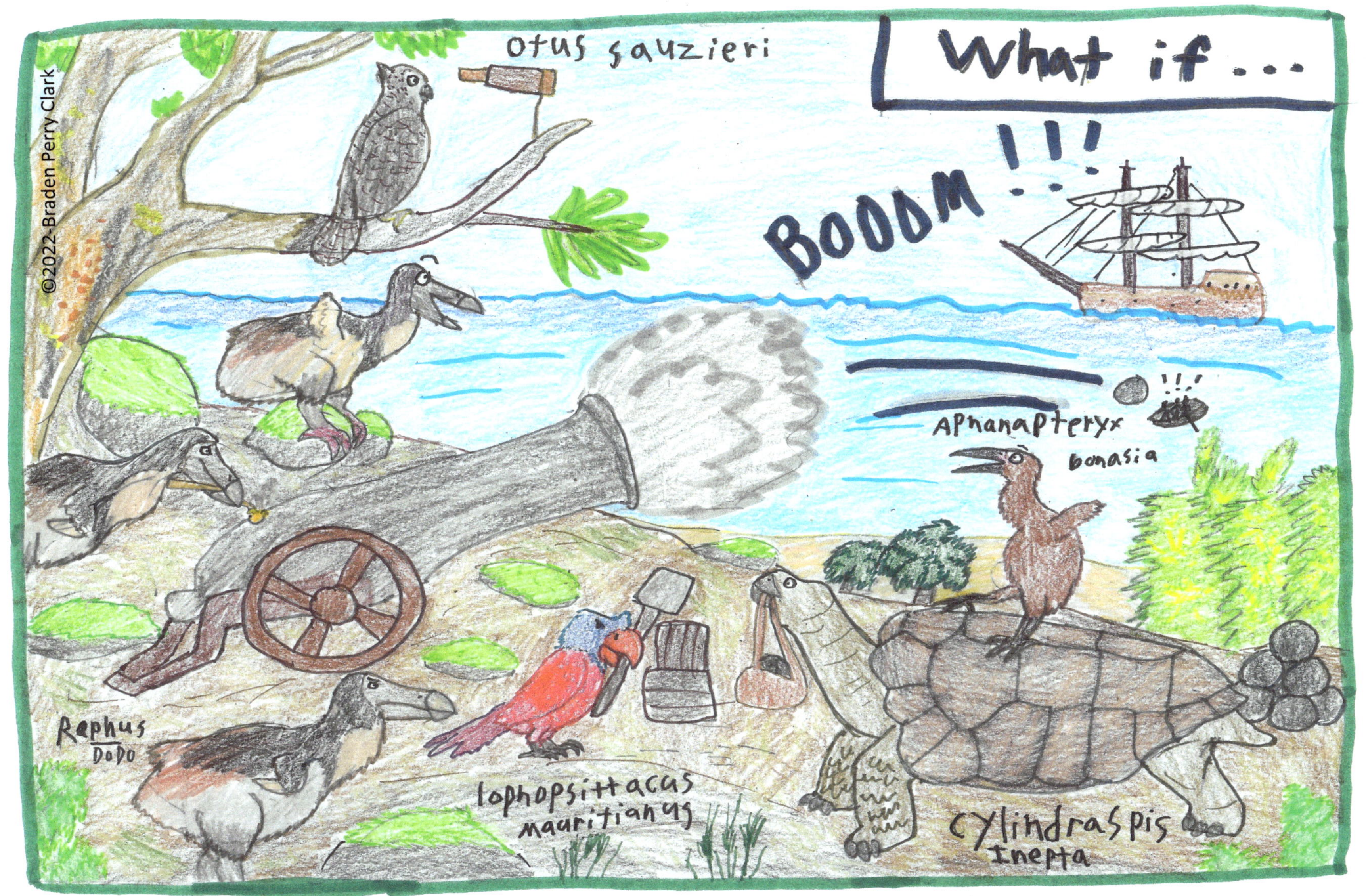

Mauritius Defense: What if the animals of Mauritius Island could defend themselves from outsiders using weapons? Mauritius Island was home to the famous Dodo along with many other unique creatures. When Europeans arrived on the island, they brought with them trouble including new invasive species, overhunting, and habitat loss which caused the extinction of some of the animals. It's a shame that the island wasn't discovered during the 20th century when humans are a little smarter in this regard.

Bulldog Fish: Xiphactinus is a large (16.7 feet long) predatory fish that lived in the Late Cretaceous Period (112-66 million years ago). It had an upturned lower jaw that resembles today's bulldog. This picture shows them chasing their favorite prey - Gillicus - which is a close relative. This scene is based on the fact that the fossilized remains of Gillicus were found in the fossil of Xiphactinus.

Achoo: If you would have visited Australia in the early Cretaceous Period, you would need to be on the lookout for the occasional sneezing Muttaburrasaurus. The Kunbarrasaurus got lucky because he was nearly hit by a snot bomb!

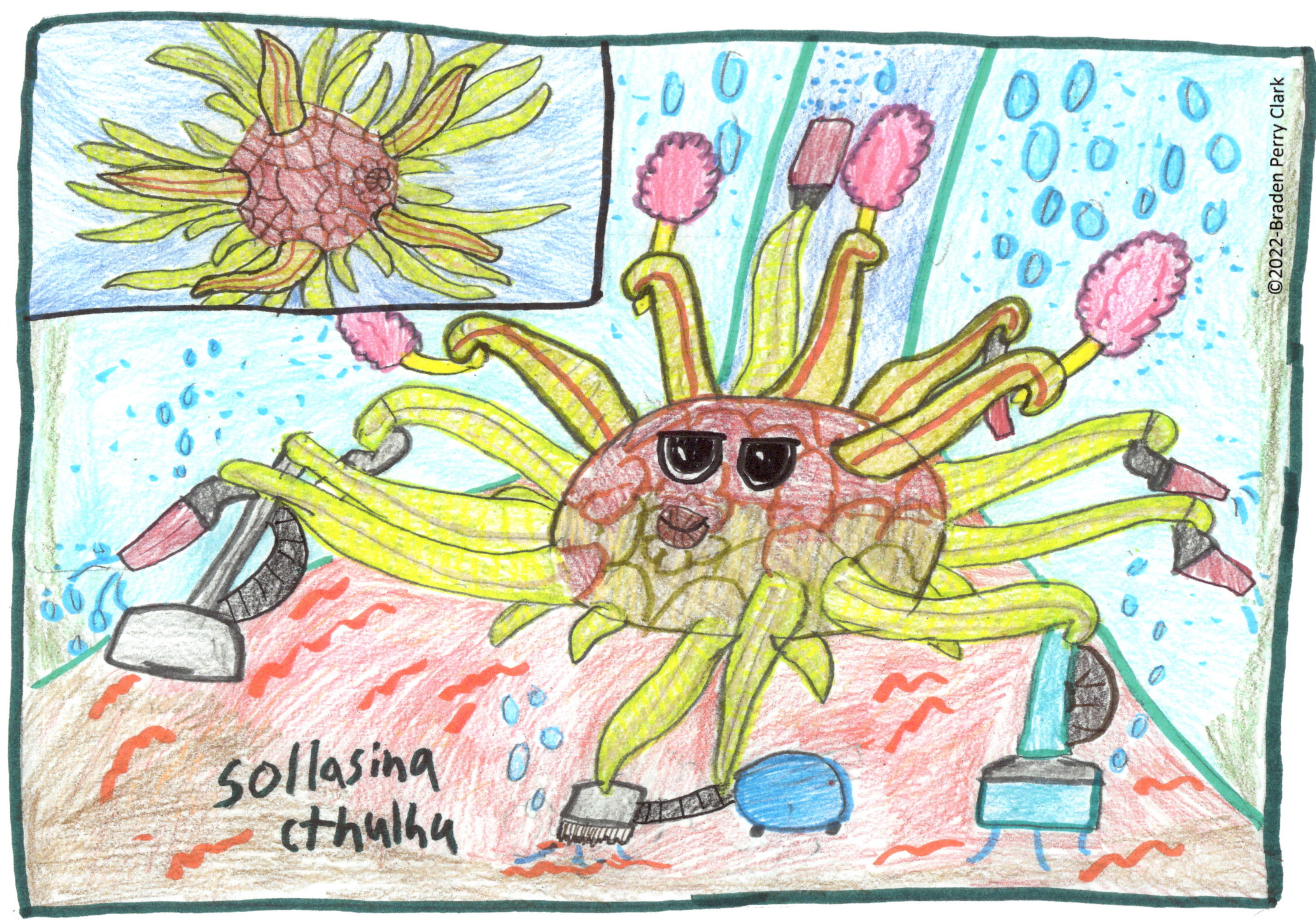

Multi-Armed Wonder: The little fella is Sollasina Cthulhu, a type of prehistoric echinoderm that lived in the Coal Brookdale Formation in England during the Silurian Period. It probably used its arms to sift through the sediment of the ocean floor to look for food. In this drawing, I imagined it used its arms to clean its house! My mother wants it to help with chores! Unfortunately, it's extinct and restricted to water!

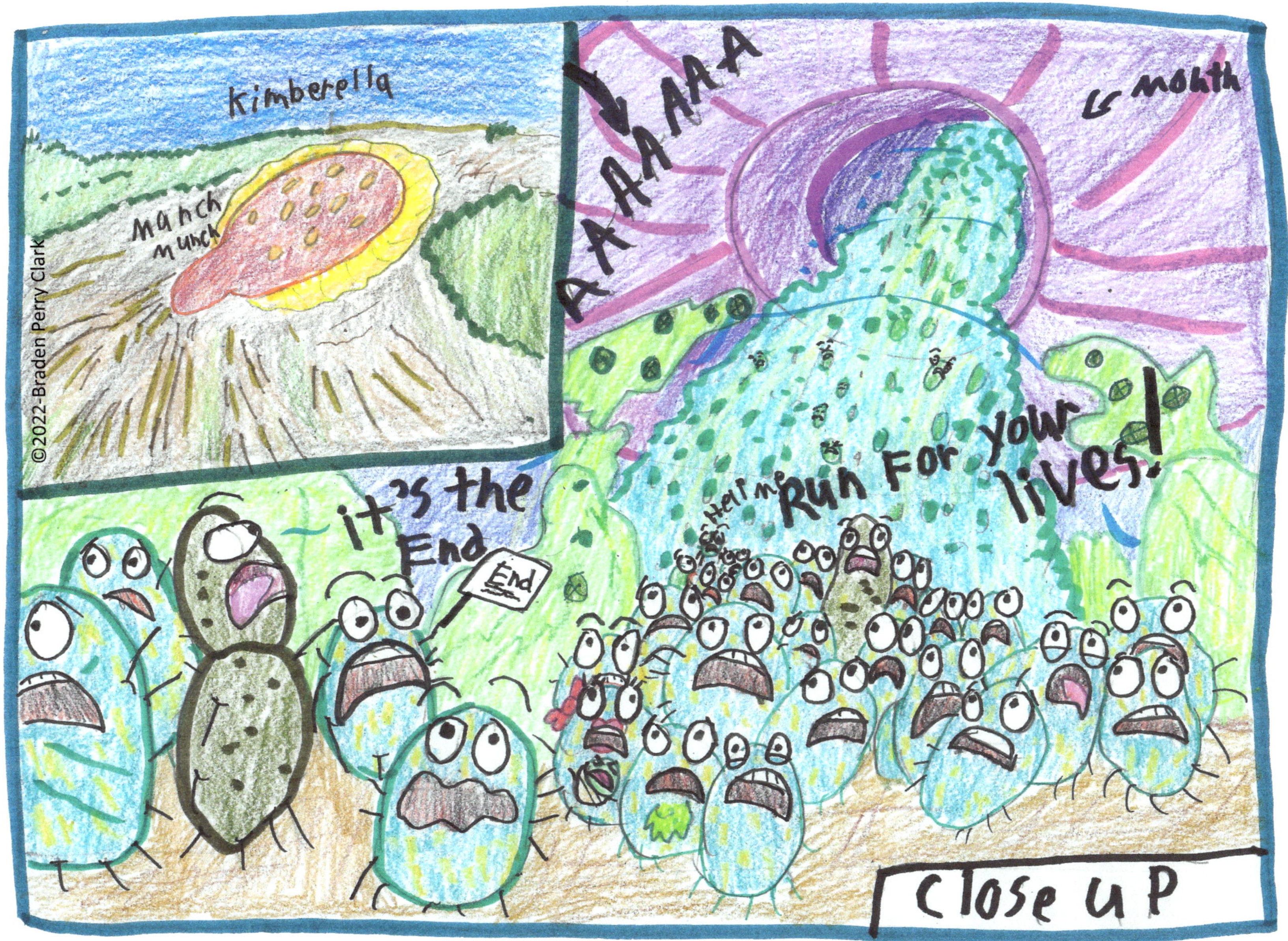

Bacterial Apocalypse: This drawing takes place during the Ediacaran period. While it looks like a scene from an apocalypse movie, this is actually the close-up of a mouth of a slug-like organism called the Kimberella. It is raking up bacteria off the ocean floor. During this period, multi-celled organisms, like this one, could only feed on algae and bacteria.

Blackbelt: A pair of Suskityrannus were about to attack a Nothronychus but then they witnessed its karate moves and its huge claws and had second thoughts!

©2021-Braden Perry Clark

Sabertooth Boxing: When North and South America connected, some animals migrated to new areas. Some are successful, but some - like this poor Thylacosmilus a Sabertooth Marsupial - couldn't handle the competition and were replaced by predators like Smilodon.

Hermit Crab Shell: This drawing is based on a real life family event. "Paul" the hermit crab is on a trip to visit his sister. Unfortunately, his wife packed the wrong shell and found out too late that it was an older shell that was supposed to be donated to the thrift store. So, he has to buy a new shell before he can go out.

Golfer's Nightmare: Just another problem at the golf course! The animals are on the Island of Malta. The elephant and hippo are pygmies while the swan is a giant. Imagine the mess!

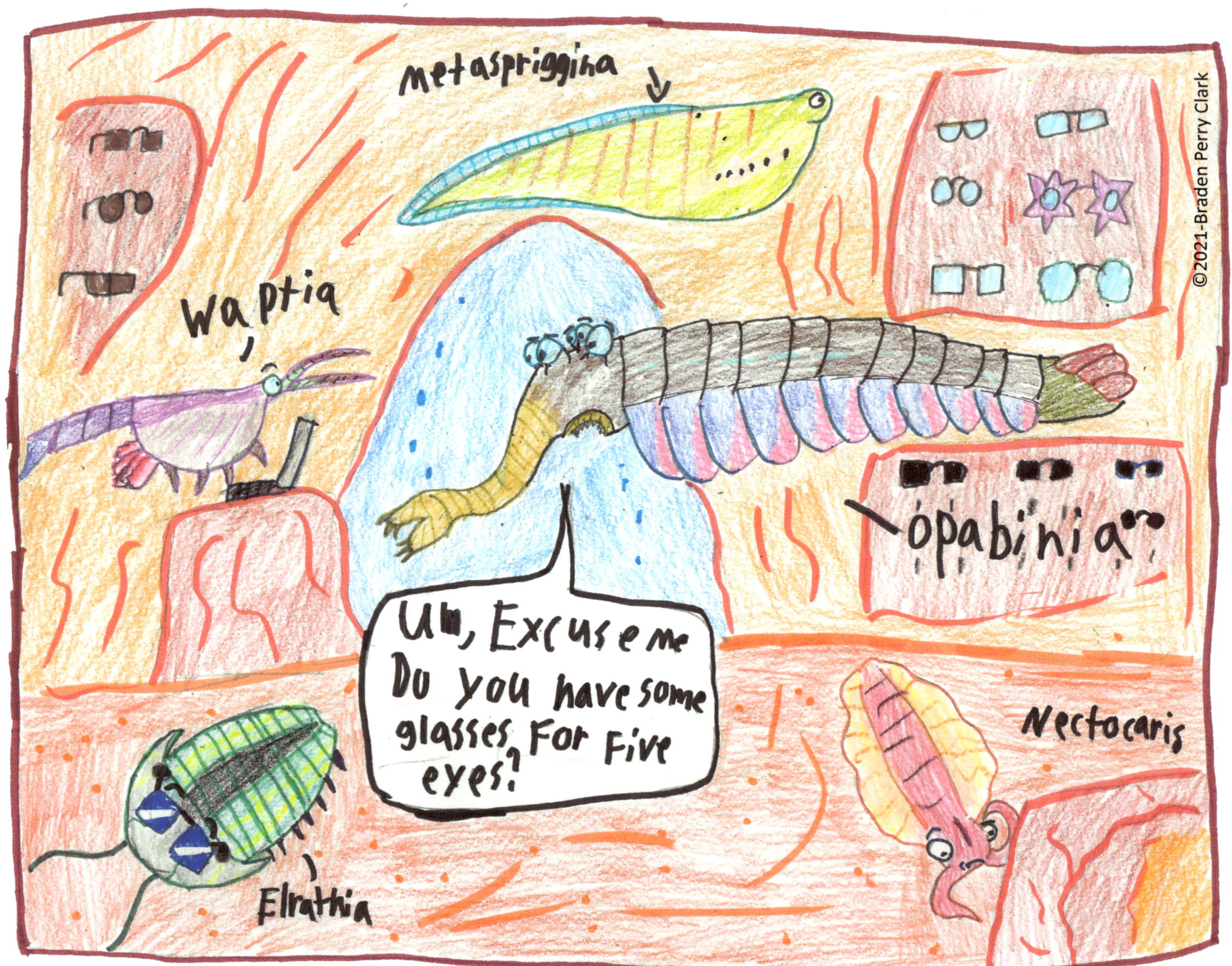

5 Eyed Problem: This is an Opabinia inside of a sponge that just happens to have an eyewear shop. He is looking for eyewear for his five eyes! However, he is having trouble finding something that fits!

Windy Day: Dimetrodon found himself in a windstorm. Since he has a sail on his back, it's a challenge to stay grounded. Dimetrodon is part of a group called Synapsids. They lived way before the dinosaurs.

Belly Flop: This is a fun imaginary scene from the Early Permian period in North America. It depicts a fun-loving Cotylorhynchus jumping into the pond to attempt a belly flop. He's trying to impress his friends, but instead, he lands on them by accident and almost squashes them. Note: they are not dinosaurs, they are Synapsids.

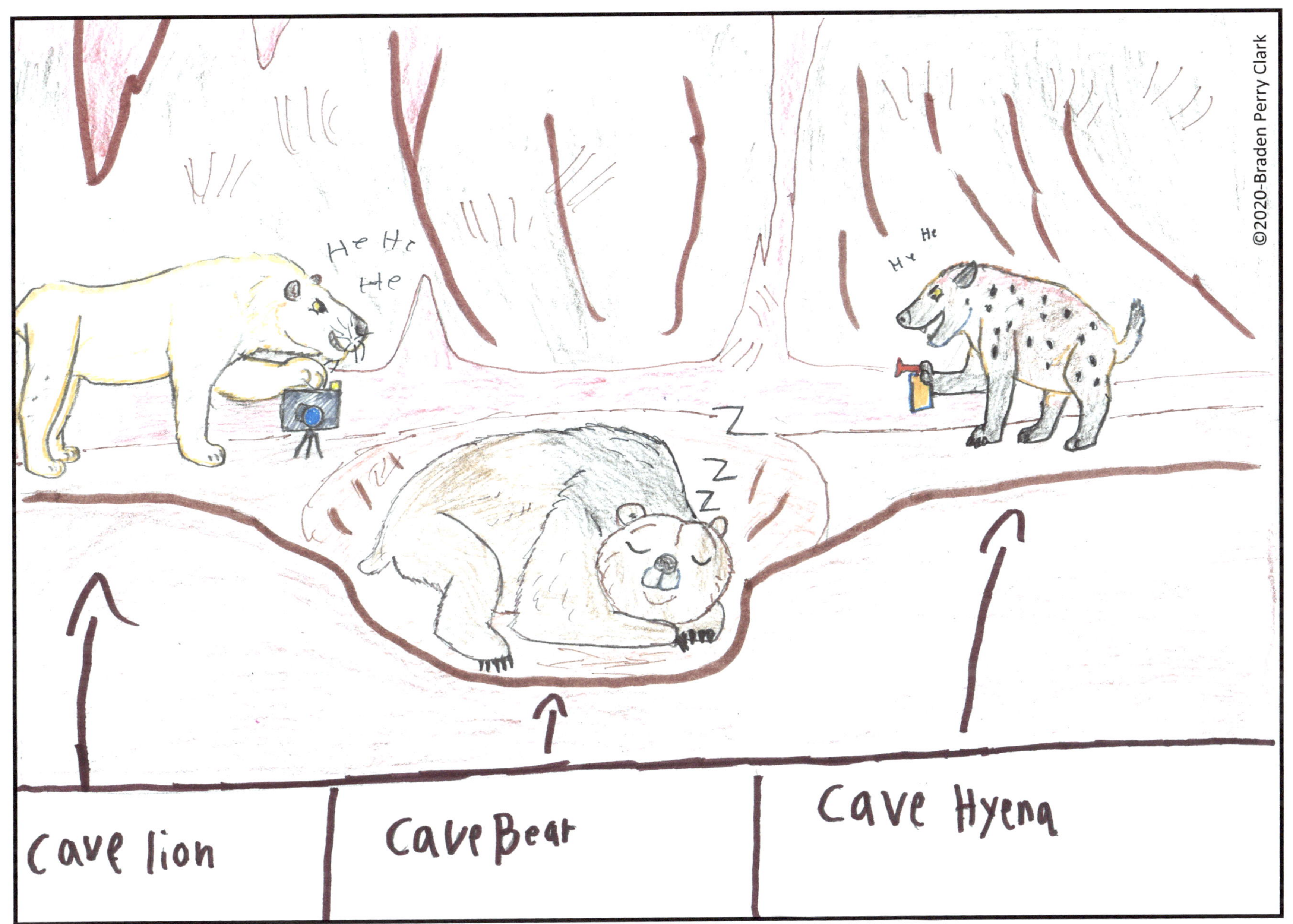

The Prank in the Cave: During the Ice Age in Europe, Cave Lion and Hyena are setting up a prank on Cave Bear. However, you can probably guess that any bear would not like being woken up -especially with an air horn!

Football: For football lovers, this drawing was based on the fact that a mammoth and other ice age animal fossils were discovered while they were remodeling Reser Stadium at Oregon State University. This drawing shows a mammoth charging at the Sabertooth cats.

Too Young to Get Eaten: This drawing is based on a real-life fossil leg bone of an ancient giant ground sloth. The fossil had bite marks from a youthful Purassaurus (Caiman) preserved in it.

Lava Dare: During the late Cretaceous Period, parts of India were covered in a volcanic landscape known as the Deccan Traps. Life still finds a way to live in extreme conditions. Fossils of Isisauras', one of the few Sauropods from India, were found in this area. They used the traps as nesting grounds. In this drawing, I imagined that the youth - or the stupid Sauropods - would dare each other to see who could keep their tail in the lava the longest!

Cold-Blooded Answer: During the Permian-Triassic extinction event, also known as the Great Dying, some species suffered while others like this Archosaurus survived and thrived!

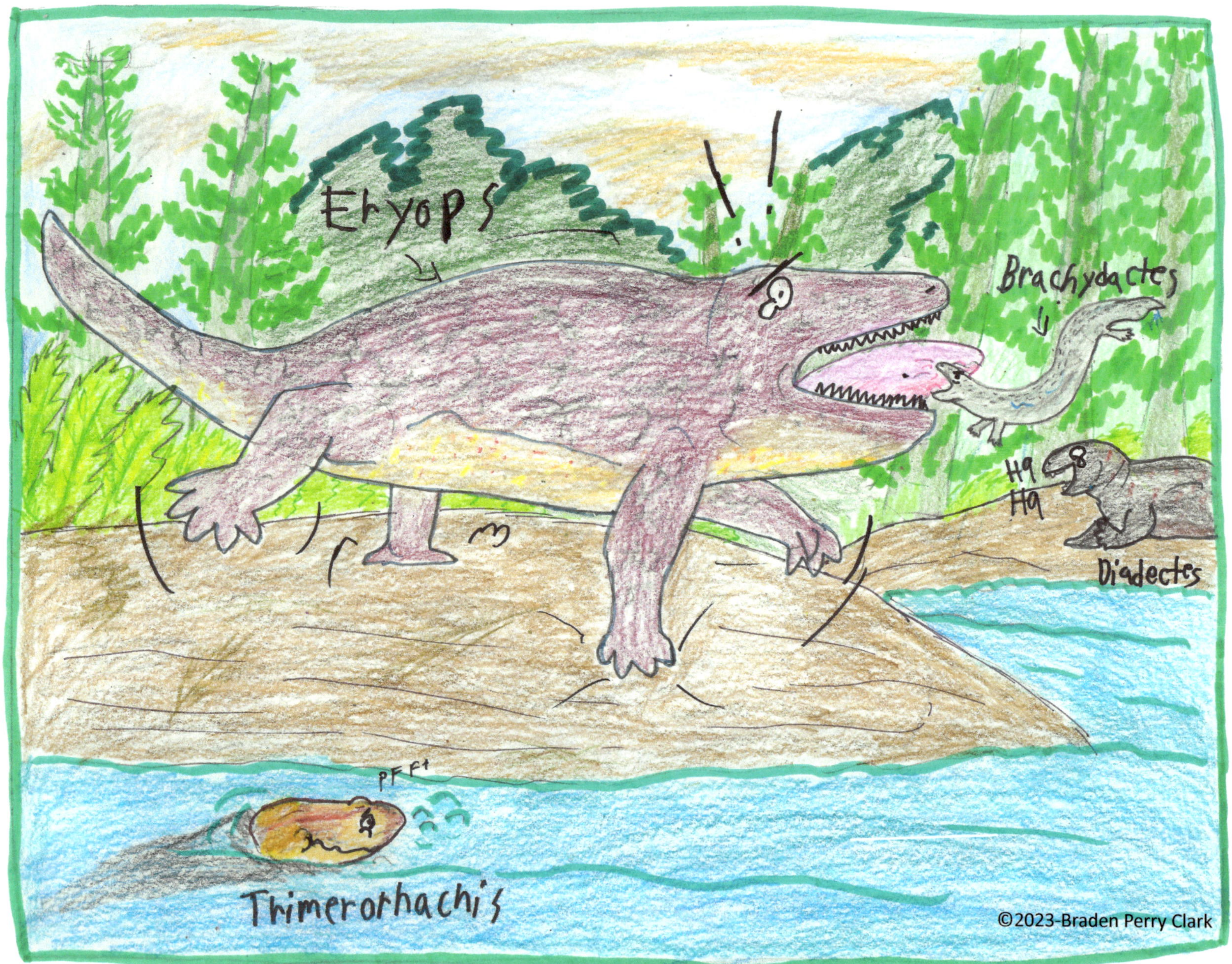

Feisty Biter: Early Permian Period - Texas - An Eryops, a type of amphibian was on the hunt when it spotted a Brachydectes and tried to eat it. The little guy didn't feel like being lunch, so it bit him back!!! The Eryops was very embarrassed by this experience, however, it was a source of amusement for locals hanging out at the swamp.

Wasps: This poor fellow here is an American Mastodon. He accidentally disturbed a nest of wasps while foraging in the Honey Locust Tree. Bees and wasps are feared by every Mastodon and even today's elephants because they fly into their trunks and sting them.

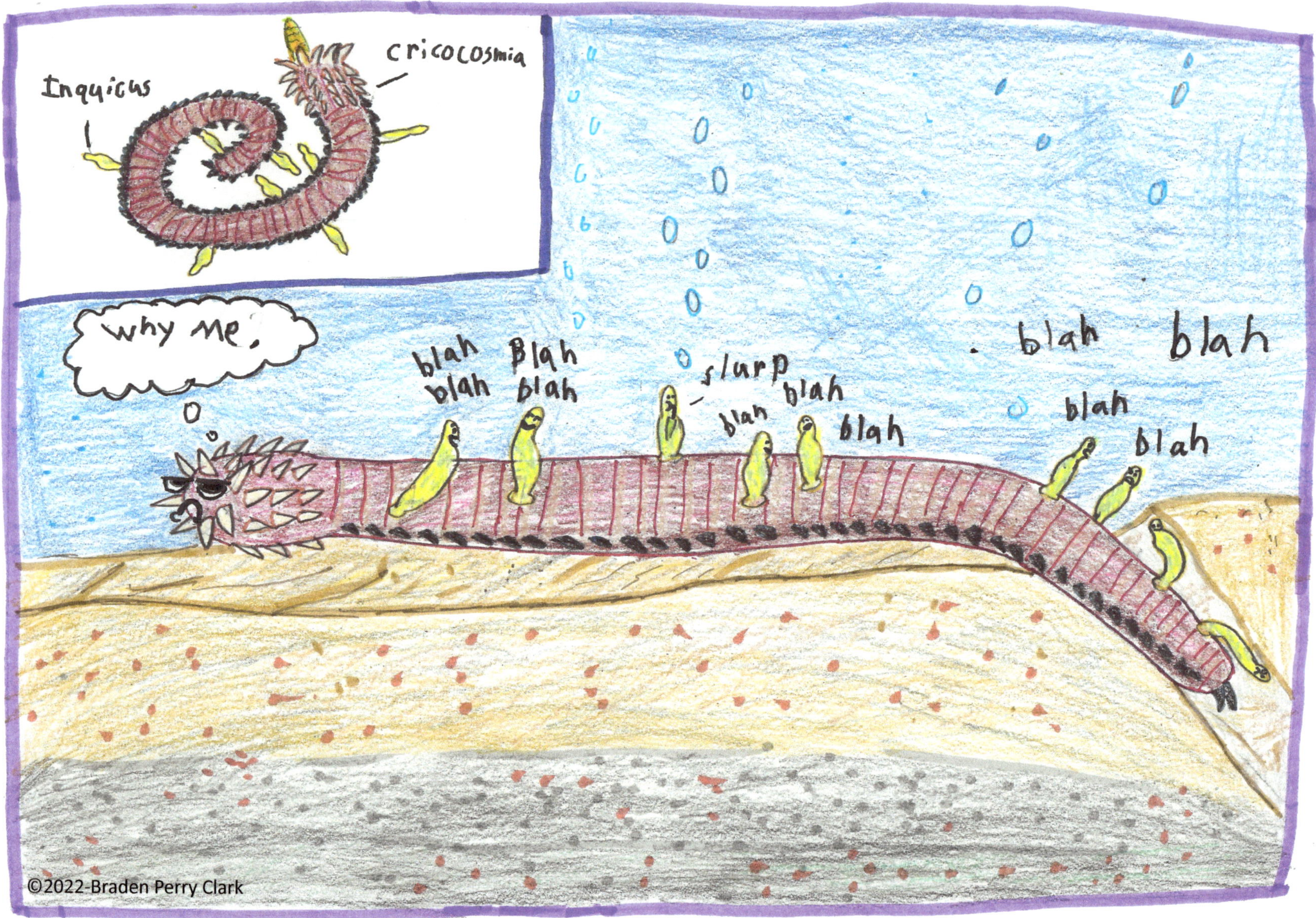

Freeloaders: Willy the Worm also known as Cricocosmia lived in China during the early Cambrian Period. He lives a simple life of eating and burrowing; however, he often picks up some unwanted Inquicus hitchhikers. They attach themselves to the worm to get around because they are a bunch of lazy invertebrates. While they don't intentionally harm the worm, they make life harder and slower than usual. Note: the worm is not wearing sunglasses. These are cartoon eyes!

Jack-O-Bird Prank: Some ancient bats (Palaeochiropteryx) are following some signs promising free food - which for them would have been moths or caddisflies. However, when they arrive at the spot, they get scared half to death! A bird of prey pops out of the tree stump! It turns out, that it's a fake prop set up by their joking friend!

All Thumbs: Once upon a time in Jurassic China in the Tiaojishan Formation, there lived a unique pterosaur named Kunpengopterus. It is unique because it has opposable thumbs. That feature probably helped it climb up tree branches or to hold stuff. I imagined it trying to taking a test. Compared to the Jenolopterus, it is acing the class!

Swiped Away: This drawing takes place in the Jiufotang Formation in China during the early Cretaceous Period. Back Story: A Shenzhoupterus, the one on the right side, caught a fish. But before he could eat it, another competitor came along and tried to take it from him. They were in a tug of war with the fish for a while until the Guidraco swooped in and snatched it from their beaks leaving them both dumbfounded! Take my advice, find a reclusive spot to eat your meal!

Unknown – Best Picture: This is my finest creation. This imaginary monster is the most dangerous of my imaginary monsters. It has every plant, fungi, microorganism, and animal in it - both alive and extinct! It can shift into multiple forms. Its origins are unknown, but it could be directly linked to the earth itself. It can become as big as a building or as small as the smallest cell. It can never die. No matter how much you do to it - chop it up or burn it - it can grow new parts. If it loses a body part, that body part could grow a new complete body. It is both boy and a girl and has regenerative capabilities. It can transform other creatures into other forms - including its favorite target - humans.

Fire Giant: This is a creature from my imagination: The FIRE GIANT an ancient being that is as old as the earth itself. Normally, it slumbers underground until it wakes up and rises to the surface. When that happens, devastation follows! Huge wildfires and volcanic eruptions spew across the land where he treads destroying everything and everyone!

The Holidays: I drew this picture because Mom wanted a holiday section in my book!

The Interruption: While expanding his home, the Cistecephalus digs into the living room of his neighbors the Diictedons. Oops! P.S. These creatures are not dinosaurs. Happy Valentines' Day!

St. Patrick's Day Snake! Fast Fact: Scientists have determined the coloration of a Colubrid Snake that lived in Spain during the Miocene Period 10 million years ago. Using new technology to scan the fossil, scientists found pigments in the fossil. Shown here, the snake is being depicted as a leprechaun who is protecting its pot of gold.

Leprechaunosaurus in Hiding: While minding his own business, the Leprechanunosaurus is spotted by some hunters and is forced to hide in a burrow with his gold. He is hoping he is protected by the hidden door and that the hunters will go away.

The Fooled Shark: Happy April Fools' Day! This drawing shows a Megalodon fooled by a piece of wood that is sculpted like a whale. The two pranksters are a small whale and a seal. This happened before the Ice Age.

May the 4th Be with You: This fellow is a Machimosaurus Rex a Teleosauroid - a type of crocodile from Tunisia whose fossils were discovered nearby the region where the filming for the Planet Tatooine from Star Wars was filmed. I am a big Star Wars fan so this drawing combines two of my interests.

Happy Cinco de Mayo Day! The pinatas in this drawing are based on real-life prehistoric animals from Mexico! The pinatas see the kids as monsters!

Happy Mother's Day! This drawing is of a Materpiscis after it gave birth to a baby boy. It is based on a fossil of the Placoderm fish that showed the baby with its umbilical cord still preserved! The fossil dated back to the late Devonian Period from the Gogo Formation in western Australia. This scene is obviously fictional. Fish do not go to hospitals to give birth. Note from Mom: I'm truly blessed to be a mother and to have Braden as my son. I love his sense of humor and he amazes me with his knowledge!

Father's Day: This drawing is a gift for my dad who has been an "Eager Beaver" doing a lot of home improvement projects lately.

Bumper Boats: I drew this picture of a Cynognathus using a water gun that didn't have much arc, so it was able to get close to the Kannemeyeria and soak him. These two are both Therapsids. In case your wondering, it's the Cynognathus' first time and he's loving it! This picture is based on true events.

Same Creatures Different Setting: This is a little meme about what happens to my art during the month of October! I am featuring a T-Rex and an Edmontosaurus in the drawing.

Halloween Night: Organism X (the orange creature) is trick or treating along with its offspring. These are the imaginary creatures that I have previously drawn. For Halloween it "shape shifted" into a pumpkin-like creature.

S'Moreasaurus: This drawing is based on an Aetosaur, a prehistoric reptile - not a dinosaur, whose marshmallow body is covered in graham crackers and chocolate! I would love to see a life-size model of this creation made out of these ingredients!

Jack Frost: I imagined Jack Frost as a Woolly Mammoth enjoying his day ice skating.

Grinch Santa Costume: I drew this picture showing my interpretation of The Grinch as a dinosaur making his Santa disguise with his pal Max.

Uncomfortable Meeting: Grinchosaurus meets Santasaurus on the rooftop in Whoville!

Nutcracker: These nutcrackers are based on Placodus and Cyamodus – marine reptiles from the Triassic Period. (In real life they eat shellfish.) I used the costumes on my family's nutcrackers as the inspiration for this drawing.

Jumpy Cubs: There is a cave in Australia that dates to the Pleistocene Period. It shows walls covered in claw marks. Those claw marks are made by the cubs of Thylacoleo - also known as the "Marsupial Lion" (which is a close relative of the wombat). I had fun imagining scenarios where the cubs would literally be climbing on the walls. I put a Christmas theme in the drawing because of how modern-day cats love climbing on Christmas trees.

Christmas Morning: Merry Christmas to all the parents! This a North American Short Face Bear from the Ice Age in its den. The Mother is not ready to wake up yet to open presents! I have based this drawing on my own experience!

Merry Christmas! I imagined Santa Claus as a Woolly Mammoth doing his yearly deliveries.

Too Big-Too Small: Two different prehistoric animals got the wrong presents. Cotylorhynchus has a small head compared to its body. The opposite is true with Erythrosuchus. I hope they can return the shirts that don't fit!

Billy in the Kitchen: I collaborated with my younger sister, Camille, on this drawing. I drew the Paraceratherium (an extinct giant hornless rhinocerotoids) who has his eye on the desserts!

Two Artists: My sister Camille drew the right side of the drawing while she was on a church mission. She mailed it to me and I came up with the idea of two missionary dinosaurs (a Nothronychus - top; Suskityrannus-bottom) doing a door approach. Elder Nothro has very long, sharp claws that always rip the Book of Mormon. His companion doesn't have very charitable thoughts. Forgive them, they are new at this! When Camille returned home from her mission to Germany, we colored the picture together.

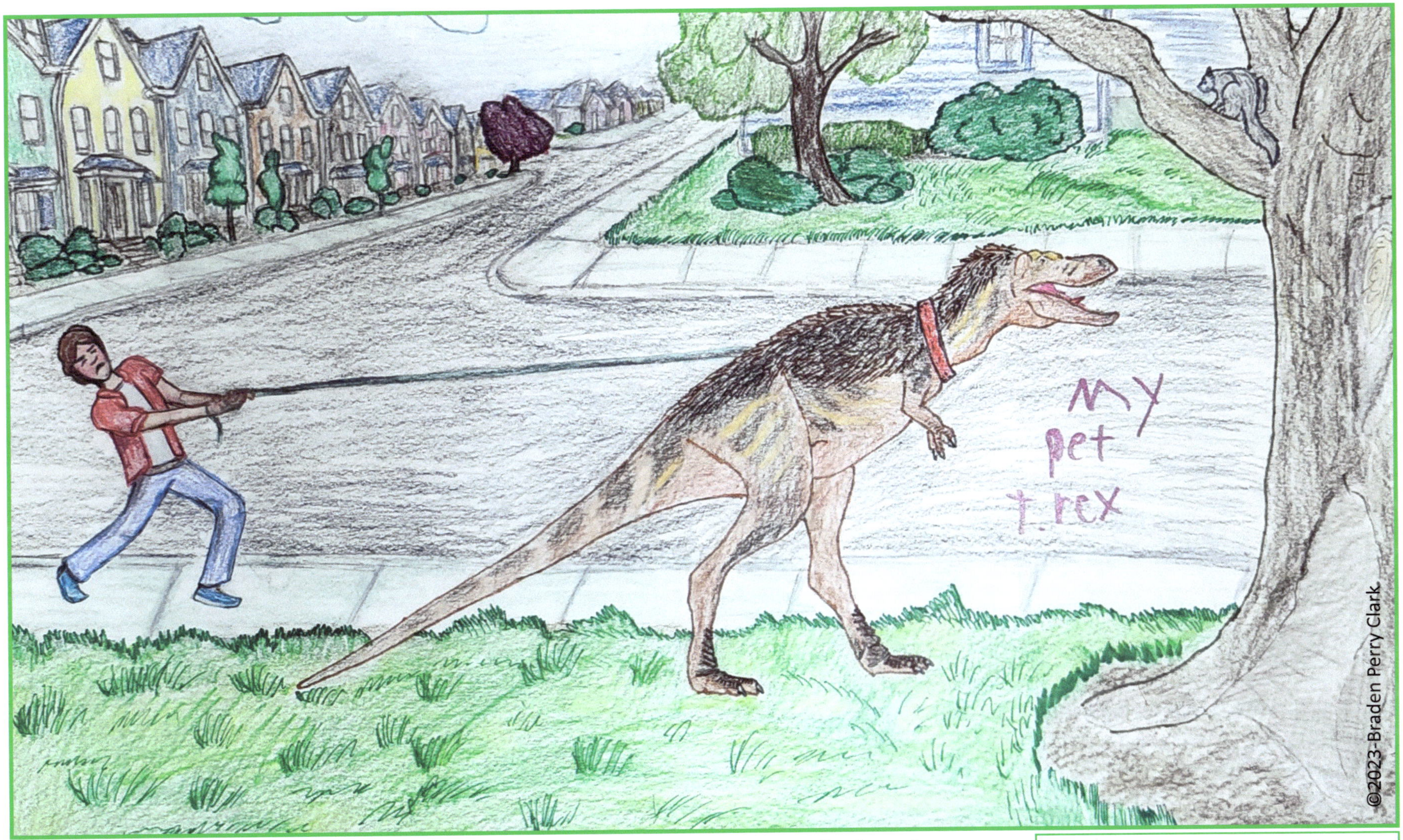

Up a Tree: I drew the juvenile T-Rex and the tree with the cat in its branches. Camille drew the guy and the background. The inspiration for this drawing came after taking the neighbor's dog on a walk! Another fun collaboration. This is an updated idea from one of Braden's first drawings seen at the right.

I hope you enjoyed my book!

- Braden